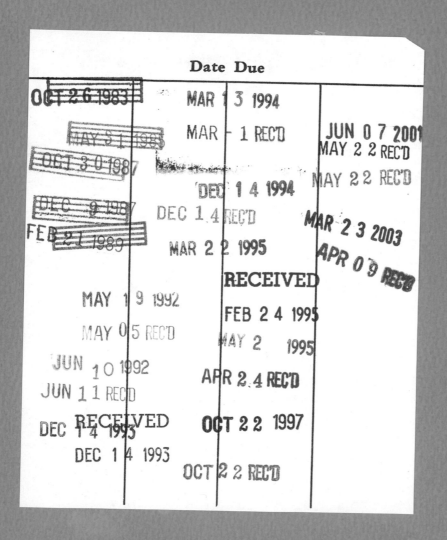

KORNEI CHUKOVSKY

THE TELEPHONE

adapted from the Russian by
William Jay Smith

in collaboration with Max Hayward

ILLUSTRATED BY BLAIR LENT

DELACORTE PRESS / SEYMOUR LAWRENCE

Library of Congress Cataloging in Publication Data

Smith, William Jay, 1918–
    The telephone.

    SUMMARY: The author's phone is rung constantly
by animals calling about their problems.
    [1. Animals—Fiction. 2. Stories in rhyme]
I. Chukovskiĭ, Korneĭ Ivanovich, 1882-1969.
Telefon. English. 1977. II. Hayward, Max, joint
author. III. Lent, Blair. IV. Title.
PZ8.3.S6712Te   [E]   75-32921

ISBN: 0-440-08532-2
ISBN: 0-440-06040-0 lib. bdg.

The telephone rang.
"Hello! Who's there?"
"The Polar Bear."
"What do you want?"
"I'm calling for the Elephant."
"What does *he* want?"
"He wants a little
Peanut brittle."

"Peanut brittle!…And for whom?"
"It's for his little
Elephant sons."
"How much does he want?"
"Oh, five or six tons.
Right now that's all
That they can manage—they're quite small."

The telephone rang. The Crocodile
Said, with a tear:
"My dearest dear,
We don't need umbrellas or mackintoshes;
My wife and baby need new galoshes;
Send us some, please!"
"Wait—wasn't it you
Who just last week ordered two
Pairs of beautiful brand-new galoshes?"

"Oh, those that came last week—they
Got gobbled up right away;
And we just can't wait—
For supper tonight
We'd like to sprinkle on our goulashes
One or two dozen delicious galoshes!"

The telephone rang. The Turtle Doves
Said: "Send us, please, some long white gloves!"

It rang again; the Chimpanzees
Giggled: "Phone books, please!"

The telephone rang. The Grizzly Bear
Said: "Grr—Grr!"
"Stop, Bear, don't growl; don't bawl!
Just tell me what you want!"
But on he went—"Grr! Grrrrrrr!..."
Why; what for?
I couldn't make out;
I just banged down the receiver.

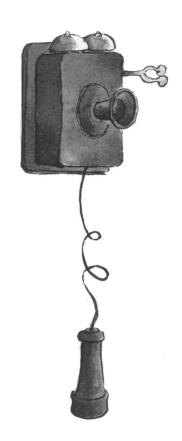

The telephone rang. The Flamingos
Said: "Rush us over a bottle of those
Little pink pills!...

 We've swallowed every frog in the lake,
And are croaking with a stomachache!"

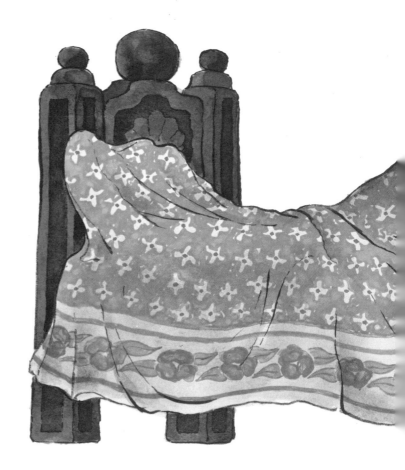

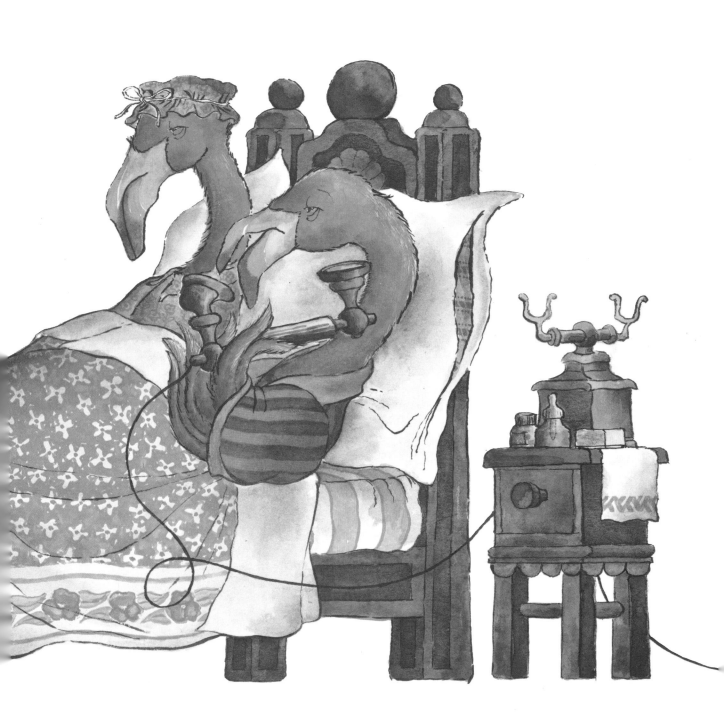

The Pig telephoned. Ivan Pigtail
Said: "Send over Nina Nightingale!
Together, I bet,
We'll sing a duet
That opera lovers will never forget!
I'll begin—"

    "No, you won't! The divine Nightingale
Accompany a Pig! Ivan Petrovich,
No!
You'd better call on Katya Crow!"

The telephone rang. The Polar Bear
Said: "Come to the aid of the Walrus, Sir!
He's about
    to choke
        on a fat
            oyster!"

And so it goes. The whole day long
The same silly song:
  Ting-a-ling!
      Ting-a-ling!
         Ting-a-ling!
A Seal telephones, and then a Gazelle,

And just now two very queer
Reindeer,
Who said: "Oh, dear, oh, dear,
Did you hear? Is it true
That the Bump-Bump Cars at the Carnival
Have all burned up?"

"Are you out of your minds, you silly Deer?
The Merry-go-round
At the Carnival still goes round,
And the Bump-Bump Cars are running, too;

You ought to go right
Out to the Carnival this very night
And buzz around in the Bump-Bump Cars
And ride the Ferris Wheel up to the stars!"

But they wouldn't listen, the silly Deer;
They just went on: "Oh, dear, oh, dear,
Did you hear? Is it true
That the Bump-Bump Cars
At the Carnival
Have all burned up?"

How wrong-headed Reindeer really are!

At five in the morning the telephone rang:
The Kangaroo
Said: "Hello, Rub-a-dub-dub,
How are you?"
Which really made me raving mad.
"I don't know any Rub-a-dub-dub,
Soapflakes! Pancakes! Bubbledy-bub!
Why don't *you*
Try calling PInhead Zero Two!..."

I haven't slept for three whole nights.
I'd really like to go to bed
And get some sleep.
But every time I lay down my head
The telephone rings.

"Who's there—Hello!"

"It's the Rhino."

"What's wrong, Rhino?"

"Terrible trouble,
Come on the double!"

"What's the matter? Why the fuss?"

"Quick. Save him…"

"Who?"

"The Hippopotamus.
He's sinking out there in that awful swamp…"

"In the swamp?"

"Yes, he's stuck."

"And if you don't come right away,
He'll drown in that terrible damp
And dismal swamp.
He'll die, he'll croak—oh, oh, oh,
Poor Hippo-
         po-
            po............."

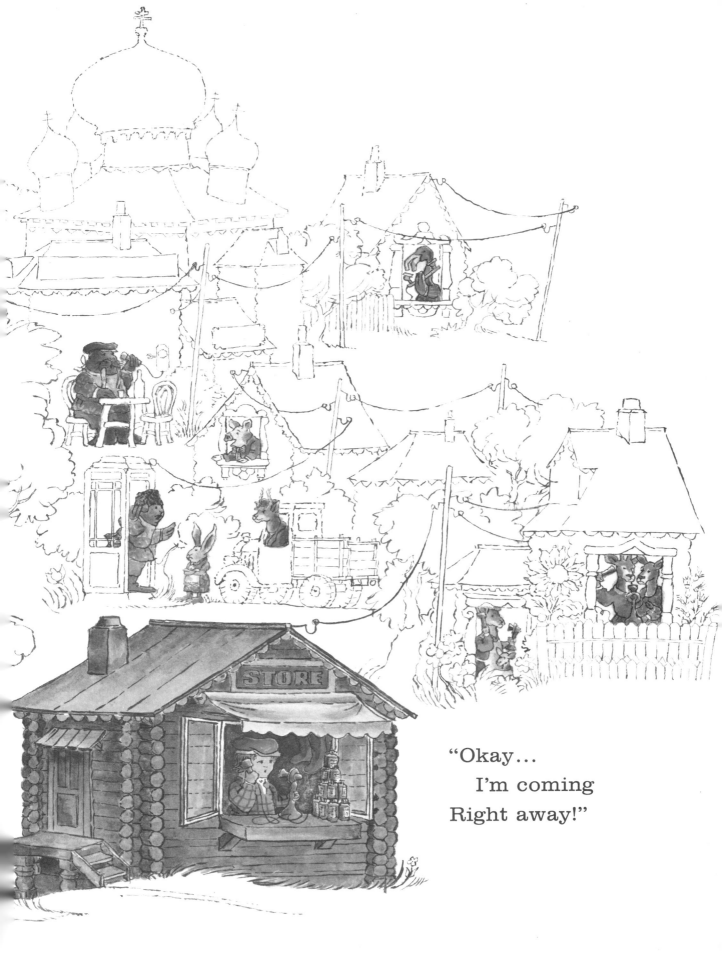

"Okay...
 I'm coming
 Right away!"

Whew! What a job! You need a truck
To help a Hippo when he's stuck!